TRUE OR FALSE?

Co-published by agreement between Shi Tu Hui and World Book, Inc.

Shi Tu Hui
Room 1807, Block 1,
#3 West Dawang Road
Chaoyang District, Beijing 100025
P.R. China

World Book, Inc.
180 North LaSalle Street
Suite 900
Chicago, Illinois 60601
USA

Library of Congress Cataloging-in-Publication Data for this volume has been applied for.

True or False? (set #4)
ISBN: 978-0-7166-5417-9 (set, hc.)

Sharks
ISBN: 978-0-7166-5426-1 (hc.)

Also available as:
ISBN: 978-0-7166-5436-0 (e-book)
ISBN: 978-0-7166-5446-9 (soft cover)

Staff

Executive Committee

President
Geoff Broderick

Vice President, Editorial
Tom Evans

Vice President, Finance
Molly Stedron

Vice President, International and Marketing
Eddy Kisman

Vice President, Technology and Operations
Jason Dole

Director, Human Resources
Bev Ecker

Editorial

Writers
Jeff De La Rosa, William D. Adams, Lauren Kelliher, Jenna Nealy, Fred Maxon

Manager, New Content
Jeff De La Rosa

Associate Manager, New Content
William D. Adams

Curriculum Designer
Caroline Davidson

Proofreader
Nathalie Strassheim

Graphics and Design

Coordinator, Design Development & Production:
Brenda Tropinski

Senior Visual Communications Designer
Melanie Bender

Senior Media Editor
Rosalia Bledsoe

Freelance Designer
Francis Lea

TRUE OR FALSE?

SHARKS

WORLD BOOK

www.worldbook.com

TRUE OR FALSE?

Sharks are fish.

TRUE!

It may sound fishy, but it's true. Sharks belong to a group of fish that also includes rays and skates.

TRUE OR FALSE?

Deadly shark attacks happen every day.

DANGER
SWIM AT
YOUR OWN
RISK!

FALSE!

Encounters with sharks are rarely fatal. Usually, shark attacks only involve one bite, and some scientists assume that sharks don't find humans particularly appetizing.

TRUE OR FALSE?

Sharks have been around for more than 450 million years.

TRUE!

As a group, sharks have survived several *mass extinctions* (die-offs), including the event that killed the dinosaurs.

TRUE OR FALSE?

Greenland sharks are known for their incredible eyesight.

Copepod

FALSE!

Greenland sharks are born with working eyes, but parasitic creatures called copepods eat away their corneas, the transparent cover of the eye. Yuck! Luckily, Greenland sharks get along just fine with their keen sense of smell.

TRUE OR FALSE?

Scientists can determine the age of a dead shark by counting the rings on its *vertebrae* (back bones).

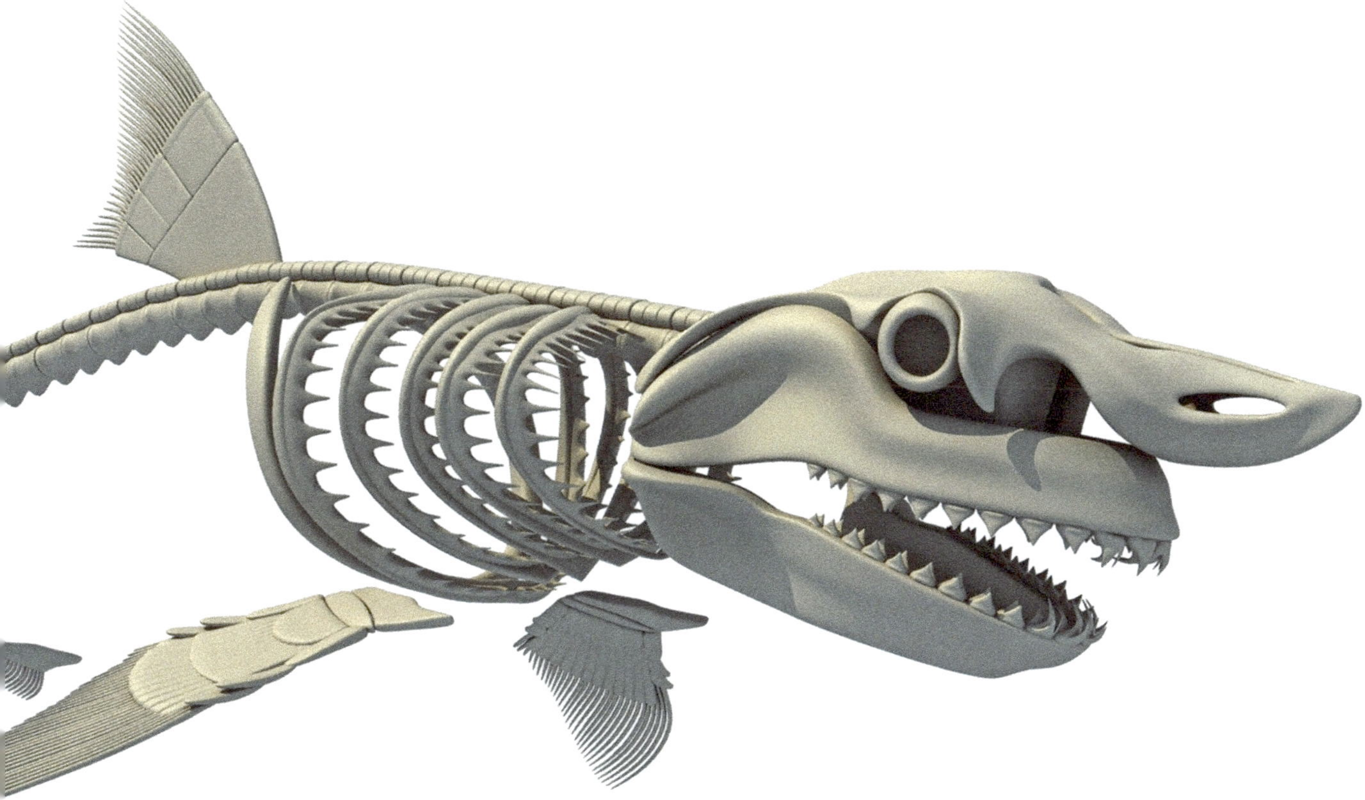

TRUE!

This method is similar to the way that scientists determine the age of a tree by counting rings in its trunk.

TRUE OR FALSE?

Sharks are known for their rigid, bony skeletons.

FALSE!

The skeletons of sharks are made up of a tough elastic substance called cartilage.

TRUE OR FALSE?

Some sharks have to keep swimming or they'll die.

Some sharks must keep moving to push oxygen-rich water over their gills. That's how sharks breathe! Other sharks can stop to rest. They use an organ called a spiracle, which gives oxygen to the brain and eyes.

Gills

TRUE OR FALSE?

A group of sharks is called a sharknado.

FALSE!

A group of sharks is called a shiver. Some shark species live alone, but others may form shivers of hundreds of individuals.

TRUE OR FALSE?

Sharks use their livers to float.

TRUE!

Sharks have oil-filled livers that help them stay buoyant without a *swim bladder*, the air-filled sac that other fish use to float.

Liver

TRUE OR FALSE?

The star of the motion picture *Jaws* (1975), nicknamed Bruce, was a trained great white shark.

Bruce was a fiberglass model of a shark that weighed over 1,200 pounds (550 kilograms) and was 25 feet (7.6 meters) long.

TRUE OR FALSE?

The cookiecutter shark is a dangerously sweet treat.

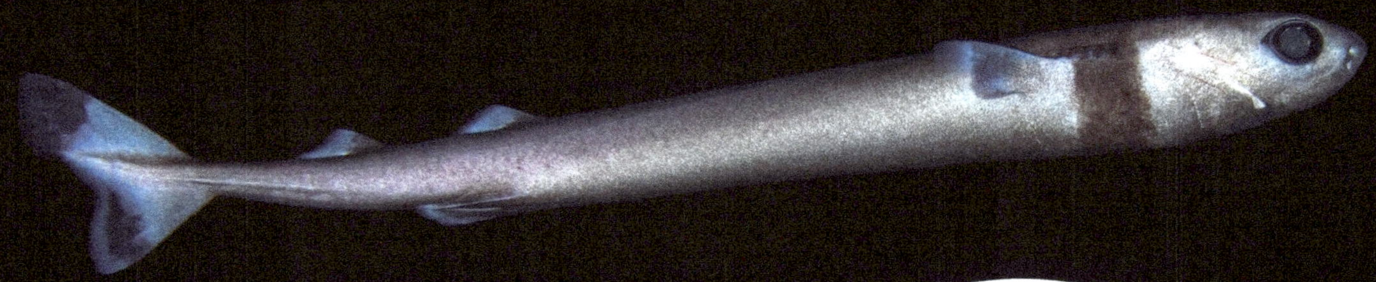

Cookiecutter shark bite mark on side of dolphin.

The cookiecutter shark is the ultimate hit-and-run predator. The 3-foot- (1-meter-) long shark is known to zip up to larger prey and snap a cookie-sized bite with its sharp teeth.

TRUE OR FALSE?

Sharks that seem angry are probably just suffering from toothaches.

FALSE!

Sharks cannot get cavities thanks to a preventive natural coating of fluoride on their teeth. Beats going to the dentist!

TRUE OR FALSE?

Sharks can have thousands of teeth.

TRUE!

Sharks have up to 50,000 teeth in one lifetime, though they only have a few hundred at a time. People, on the other hand, have to get through life with only about 52 (including baby and adult teeth).

TRUE OR FALSE?

Sharks can live for hundreds of years.

TRUE!

Scientists estimate that the Greenland shark can live at least 250 years. Some may live for over 500 years!

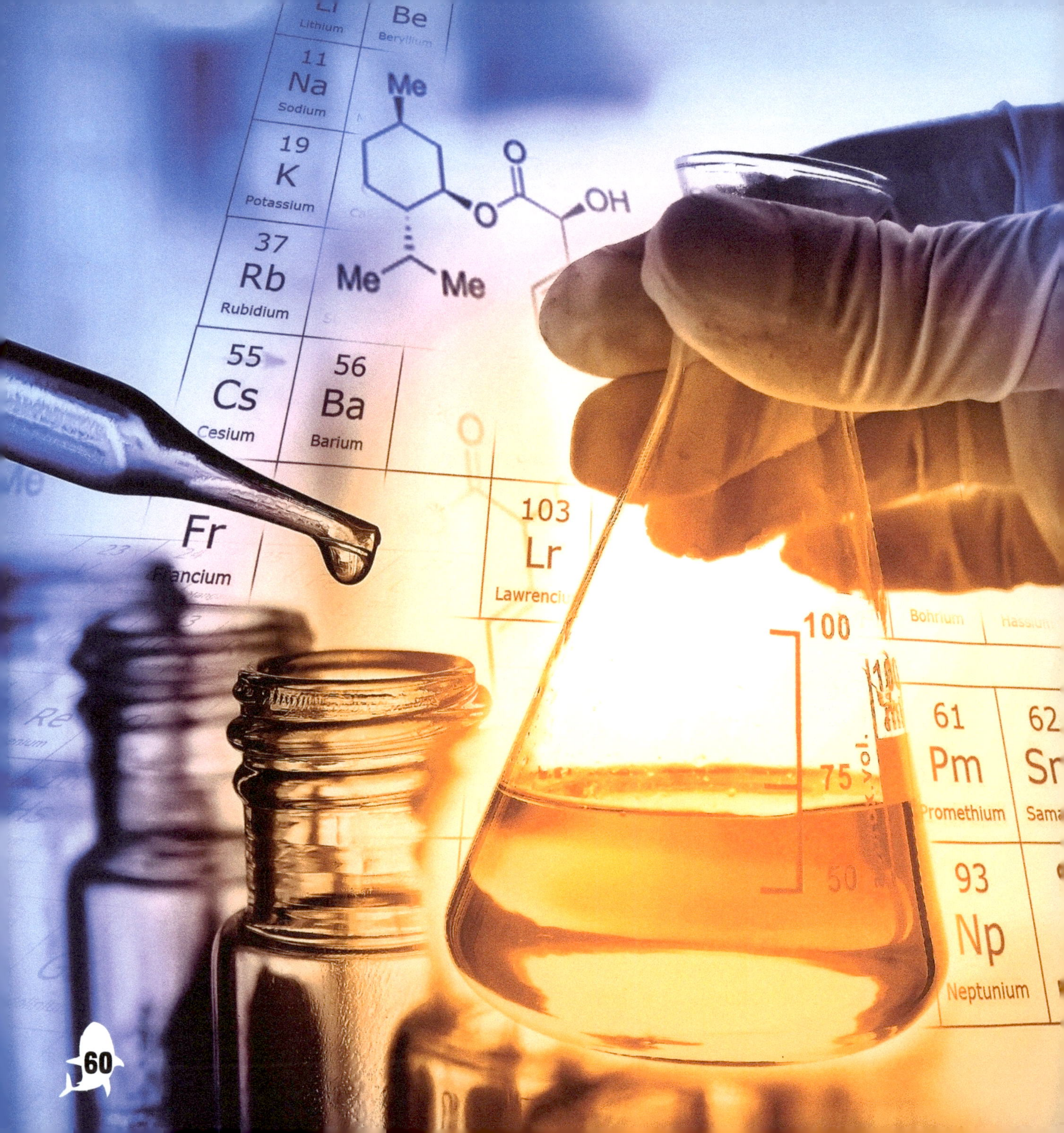
Be
Beryllium
Lithium
11
Na
Sodium
19
K
Potassium
37
Rb
Rubidium
55
Cs
Cesium
56
Ba
Barium
Fr
Me
Me
Me
O
OH
103
Lr
Bohrium
100
75
50
61
Pm
Promethium
93
Np
Neptunium
62

TRUE OR FALSE?

Squalamine, a chemical compound found in tiger sharks, is used in medicine.

TRUE!

Squalamine is used to treat bacterial infections, eye diseases, and cancer. Squalene, a different chemical compound, is an oil found in shark livers. It's used in cosmetics.

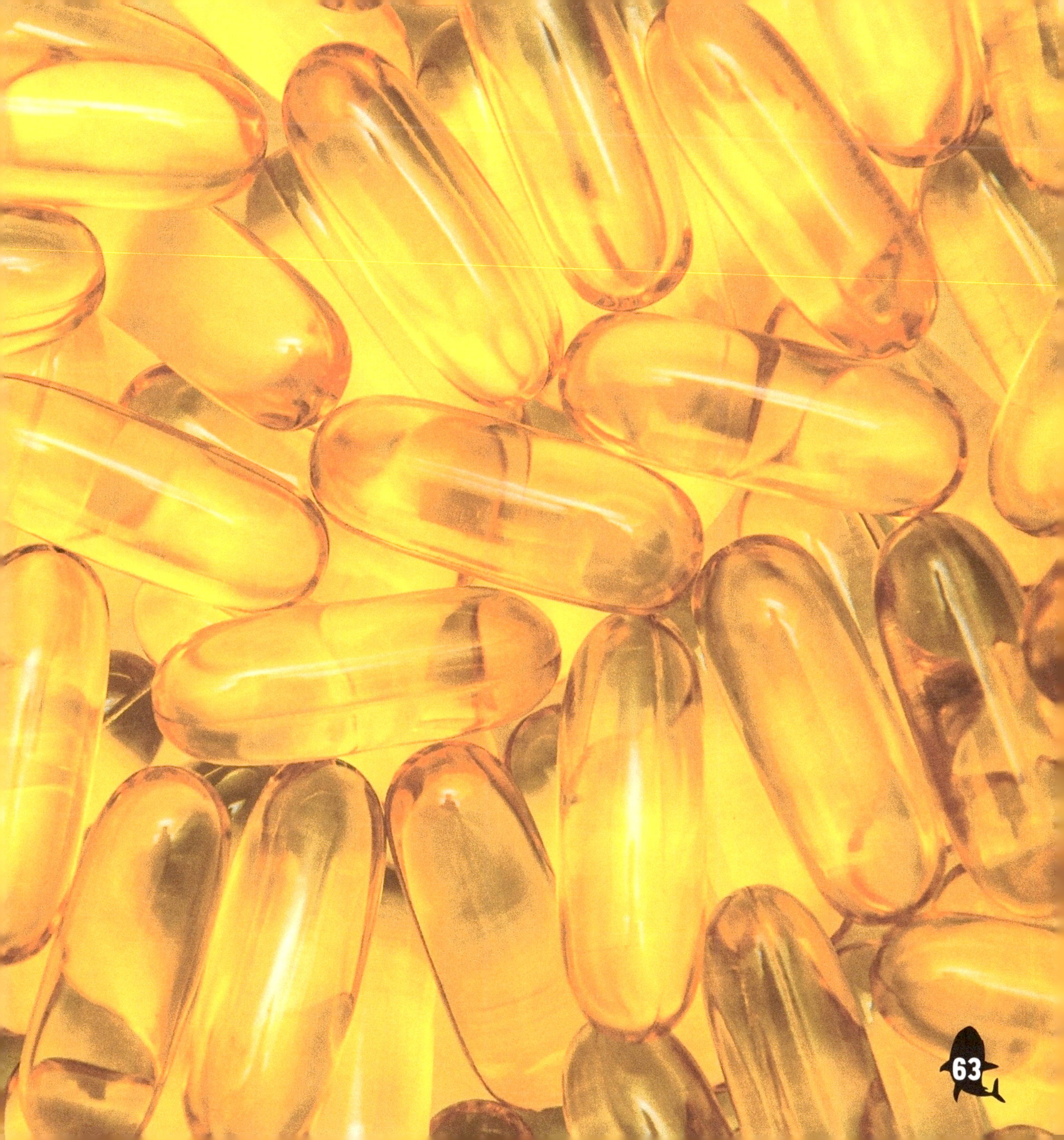

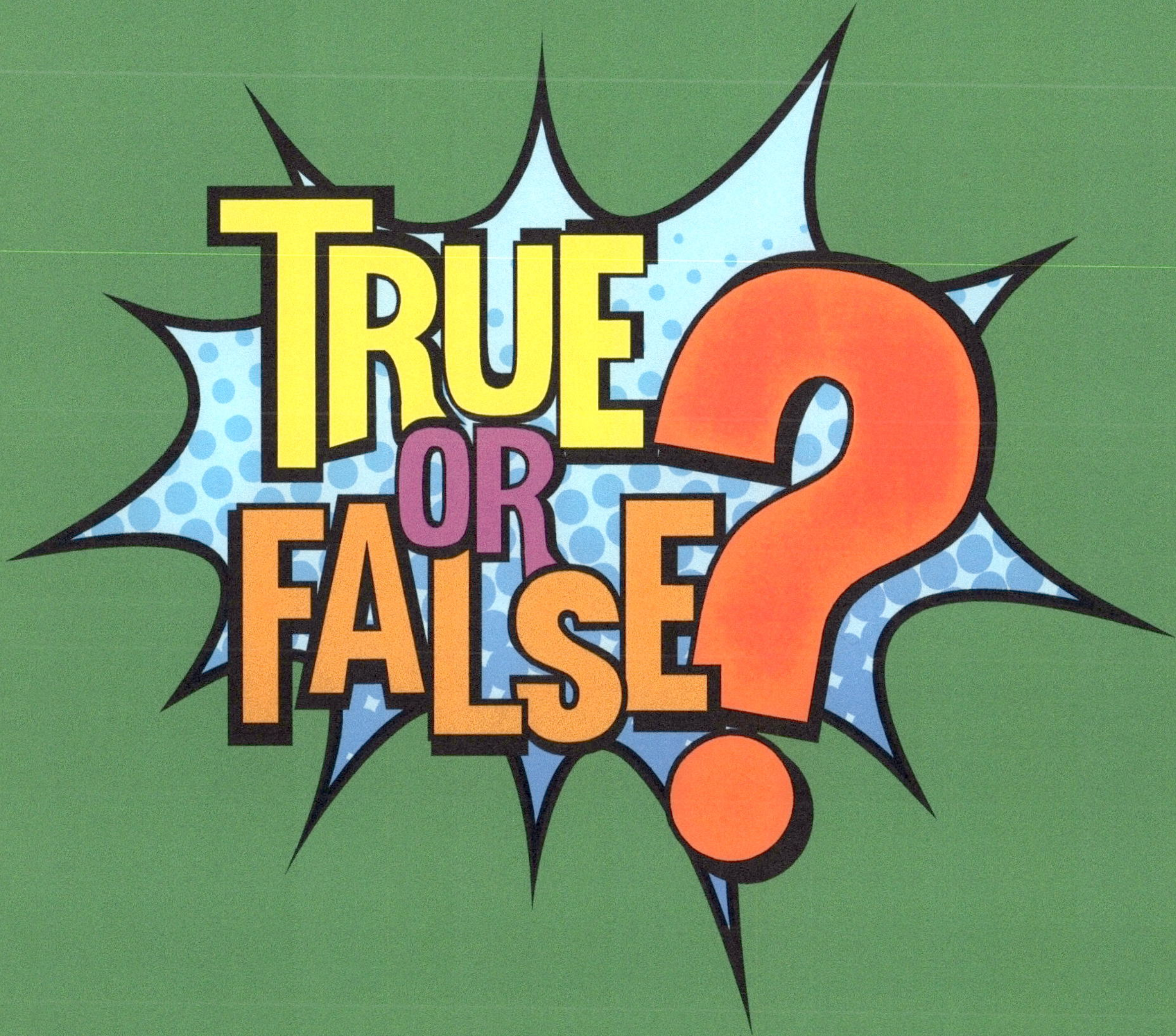

The megamouth shark—the third largest living species—was discovered by pirates in the 1700's.

FALSE!

This 16-foot- (5-meter-) long monster managed to stay hidden until 1976. Its deep ocean habitat kept it out of sight of scientists and fishing operations.

TRUE OR FALSE?

The basking shark is the largest of all sharks.

FALSE!

The basking shark is the second largest shark. It can reach up to 40 feet (12 meters) long.

The whale shark is the largest shark. It is known to reach a length of at least 46 feet (14 meters).

TRUE OR FALSE?

Baby sharks are called kittens.

FALSE!

Baby sharks are called pups. Among most species of shark, the eggs hatch inside the female, and the pups are born alive. Woof!

..zzZZ

TRUE OR FALSE?

Some sharks can take half a nap.

TRUE!

Some sleepy sharks can let half of their brain rest, while the other half stays alert.

TRUE OR FALSE?

Hammerhead sharks use their hammer-shaped heads to pound in nails.

The "hammer" of a hammerhead shark, called a cephalofoil, gives the animal souped-up senses.

With eyes widely spaced at either end of the cephalofoil, the hammerhead has excellent binocular vision. Widely spaced nostrils give the shark binocular smell, too!

TRUE OR FALSE?

The song "Baby Shark" became the first YouTube video to achieve 1 billion views.

PINKFONG
doo doo doo doo doo doo doo.

TRUE!

"Baby Shark," once a popular campfire song, *went viral* (exploded in popularity) after being recorded in 2016 by the Korean company Pinkfong!

TRUE OR FALSE?

Bull sharks often live in fresh water.

TRUE!

The bull shark is one of the few sharks that can live in fresh water as well as salt water. Young bull sharks may live in rivers and lakes. Living there enables young sharks to avoid predators.

DID YOU KNOW...

Great white sharks can **migrate thousands of miles or kilometers** each year to take advantage of the best feeding grounds.

The word shark may have come from the German word **Schurke, meaning scoundrel.** That word was first applied to people who preyed on others, such as **loan sharks!**

Sharks have a **corkscrew-shaped digestive tract.** It helps keep food flowing in one direction without using extra muscles or energy.

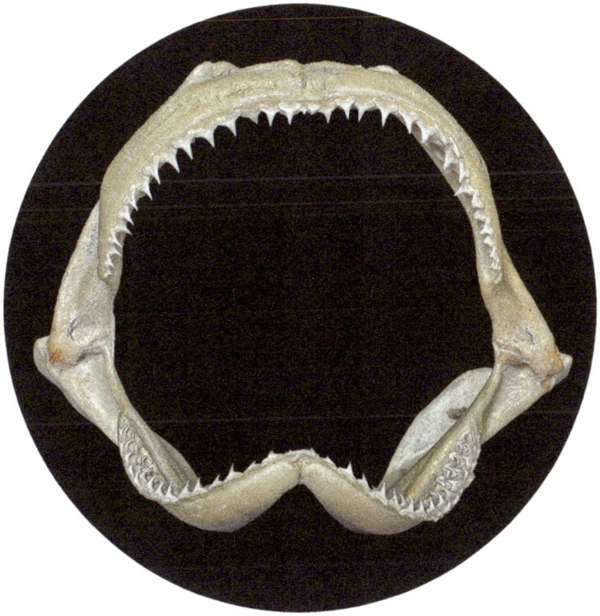

Shark jaws are not attached to their skulls, enabling both top and bottom jaws to swivel.

Sharks can be found in **every ocean on Earth.**

ENGAGE YOUR READER

GUIDED READING PROMPTS

Before Reading

- Allow readers to scan the text and discuss what they notice so far. Highlight the structure of this text and explain that the answers include both evidence and reasoning that support the claim of true or false.
- Explain the literacy skill: *Sometimes authors write a claim and then use evidence and reasoning to help make their point clear. Look for these elements as you read!*

During Reading

- Read each statement and provide time to discuss whether readers believe it to be true or false before turning the page to learn the facts.
- As you read, model how to identify the claims, evidence, and reasoning in the text. Prompt your readers to identify these features as they explore the text, too.
- Encourage readers to further discuss their learning by pausing to discuss surprising information.

After Reading

- Prompt your readers to connect, extend, and challenge their thinking about the text:
  - What will you take away from reading this text?
  - What changes in your thinking happened while reading and learning?
  - What is still challenging your thinking? What questions or wonderings do you still have?

LOOK BACK!

- Prompt readers to look back through the text to identify examples of interesting or thought-provoking claims.
- Challenge readers to explain what makes these examples so engaging.

CURRICULUM CONNECTIONS

These questions and tasks support the following English/Language Arts skills:

- Determining what a text says both explicitly and implicitly
- Citing specific evidence when drawing conclusions
- Interpreting words and phrases used in a text
- Analyzing how the structure of a text affects how it is read.

LITERACY SKILL

Authors make their claims stronger by supporting them with evidence and reasoning.

- A claim is a statement of truth.
- Evidence includes the facts or information that prove whether the claim is true.
- Reasoning includes any logical explanation that describes how the evidence supports the claim.

Example from the text: Pages 28-31

- Claim: Some sharks must keep swimming or else they will die.
- Evidence: Sharks need the oxygen from the surrounding water to survive.
- Reasoning: Some sharks' bodies are built so they can only get oxygen when water pushes over their gills. For that to happen, they must constantly move through the water.

EXTEND THROUGH WRITING

Challenge readers to create their own True/False questions and answers about sharks.

- Have readers use a trusted reference, such as www.worldbookonline.com, to research information related to sharks. Encourage readers to look for key details, fun facts, or surprising features that would make strong True or False statements.
- Give readers one notecard for each claim they research.
- Direct readers to write the claim on the front of the notecard. On the back, readers should describe why that claim is true or false using evidence and reasoning from their research.

MORE WAYS TO ENGAGE!

- Play a game! After considering each claim, have readers signify "true" with a thumb up and "false" with a thumb down. Keep score to see who knows their facts about sharks the best!
    - Develop collaboration skills by grouping readers together into teams.
- Further discuss any True/False claims that revealed readers' misconceptions. Focus the conversation on *why* they initially thought what they did and how the text helped them learn.

Acknowledgments

Cover	© C Nuisin, Shutterstock; © Khosro/Shutterstock; © Teguh Mujiono, Shutterstock
4-17	© Shutterstock
18-19	© Franco Banfi, Nature Picture Library/ Alamy Images
20-21	© 3D-Horse/Shutterstock
22-23	© Gary Corbett, Alamy Images
24-25	© Marina Mazitova, Alamy Images
26-27	© Michael Wheatley, Alamy Images
28-37	© Shutterstock
38-39	© Zaie/Shutterstock; © Panther Media GmbH, Alamy Images
40-41	© Universal PicturesDavid Pereiras
42-43	Todd Wawrychuk/Courtesy of AMPAS
44-45	© Laurentiu Iordache, Alamy Images
46-47	© Blue Planet Archive/Alamy Images; © Yunosuke/Shutterstock
48-65	© Shutterstock
66-67	© Bios Photo/Alamy Images
68-69	© Charles Hood, Alamy Images
70-83	© Shutterstock
84-87	© Pinkfong; © David Pereiras, Shutterstock
88-96	© Shutterstock

www.ingramcontent.com/pod-product-compliance
Ingram Content Group UK Ltd.
Pitfield, Milton Keynes, MK11 3LW, UK
UKHW060218061026
15625UKWH00007B/384

9 780716 654261